How Audiobooks Make You Smarter

7 Little Known Ways Audiobooks Can Boost Memory Capacity And Increase Intelligence

I0484462

Entrepreneur Publishing

COPYRIGHT

DISCLAIMER

Kindle 5 Star Books

Free Kindle 5 Star Book Club Membership

Join Other Kindle 5 Star Members Who Are Getting Private Access To Weekly Free Kindle Book Promotions

Get free Kindle books

Stay connected:

Join our Facebook group

Follow Kindle 5 Star on Twitter

Also, if you want to receive updates on Entrepreneur Publishing's new books, free promotions and Kindle countdown deals sign up to their New Release Mailing List.

Table of Contents

Introduction

Why this book?

In all these years, I have not had an opportunity to give back as much as I have gotten this year, and I want to thank you for the opportunity to explain myself.

I have a father who has been suffering with diabetes for years. This year, it has been explained to him that he may lose his eyesight, and that troubled him greatly. He has been such a visual person all his life, the thought of losing his sight terrified him.

Then, I helped him discover the one thing that he could enjoy, could still experience, that would offer him the same educational and entertainment value he had gotten through his ears instead, and that has made all the difference.

Audiobooks represent a significant business today, with all the various methods of file transmission, recording methods, and the sheer number of available titles.

This book, I hope, will offer you and yours the greatest opportunities in learning, expanding your horizons, and memory retention possible.

Using the foremost in research on how we hear, how we learn, and developing the means to further improve upon them, is precisely what this book is intended to do. Within its pages, I hope to share with you how hearing works in learning and data access and retrieval, so you can see the ways that Audiobooks can activate and amplify your ability to learn. We will be discussing the very ways that hearing works in locking down memories, showing some insight how we can speed up, and amplify those effects.

Additionally, this book will explain the means to utilize visualization and other learning strategies to increase recall speeds and actively improve retention of information, and will explore the possibilities of taking advantage of the power of the subconscious mind while listening.

Other topics we will cover include triggering the optimal outcome of our natural active, passive, and subconscious learning methods, to give our minds permission to recover information faster and more accurately. To do so, I will also have to explain some of the scientific research that his helping us unlock mysteries of the mind, such as outlining current theories about how cognition and hearing are interrelated, and what can be done to improve information intake

With the success my father has had, acclimating to a less visual, more auditory lifestyle, I believe your successes and discoveries that can be made through Audiobook learning processes outlined in this book will make for unparalleled increases in your own educational processes and information retentions matrices.

In the end, I hope this book will show you can use Audiobooks as conduits for communication, and build bridges with the larger world beyond.

Chapter 1: Hearing Is The Primary Resource For Retention

In this chapter, you will learn:

- Key parts of the ear, now they work together, and

- Understanding the auditory nervous system.

Auditory Function is Constant

As you may know, the one sense we nearly all would list as the one we would hate to lose is vision. We are so dependent on sight that we quite often fail to recall that unlike sight, the primary sense upon which we depend on the most is our hearing.

Hearing impacts every one of us, every day, in a myriad of ways. Most often, we only concentrate on the specifics we might listen for - danger, our loved ones, or beautiful sounds such as those of nature or music. But what we don't always realize are those functions that interact with the outside world.

Learning, therefore, will always have a function in the learning process. The more we can incorporate hearing in the process, the more natural that learning process becomes, because we are tying that which we do as a matter of living into that which can improve out individual lives. Thus, we create a positive and recursive learning spiral, improving for the purpose of increasing the desire to improve.

Hearing as a function of balance

While we usually don't consider it, the speed of the wind, the stability of the soil, even the proximity of perils can be reacted to if we are hearing properly. Not only the audible sounds themselves, but also the movement and pressure of the air around us provide the feedback correlation necessary between the gyroscopic nature of the inner ear, and our cognition of our surroundings While it wouldn't seem to make a direct relationship, the ability for us to learn is dependent on that balance feature, because without the ability to coordinate what we are hearing with our other sensory inputs, we literally couldn't walk and chew bubble gum at the same time.

How Sounds play into memory.

Echoic Memory

To understand why the auditory process in general, and audiobooks in particular can improve our learning processes, and improve our intellect, we first have to have an understanding of how we, as human beings, store memories. While science and medicine have yet to unlock these mysteries, a great deal can be understood if we think about our own learning process.

OPTICAL

When we look at something, we have the capacity to hold that vision, and continually receive verification of it, of the data it represents again and again. In the case of learning, we can even review it more than a singular experience. The need for speed in this process is clear; there is a lot to see, and our brains need to process quickly. Thus, memories supplant what we actually see with symbols and the interconnection between them. Retention of these long-term symbols and the interrelationship between them is called **iconic memory.** The conversion process from visual to iconic memory is rapid, and that is why what we see and what we retain are usually very small percentages of our learning; it is sheer repetition and continuous learning that embeds the information for later recall.

AUDITORY

Hearing, on the other hand, is a very different kind of mental input. Unlike sight, wherein one can continue to take in information, and even visually review such input, sounds generally are 'hearable' a single time. In the mind, therefore, a sound has to be maintained in memory, effectively, as an echo of the sound itself. Thus, when we recall a sound, a noise, or even words, our brains fundamentally recall the specific sound as a 're-play' of it. That is to say, the actual sound response is held, not as the cognizant representation (as in optic memories) but as what is now known as '**echoic memory**'. The term was coined in 1967, by Uric Neisser, who was doing research on how audial memory is retained. In essence, the brain captures a sound (almost literally as we describe audibly memorable information) fundamentally as a sound byte, that it holds and replays for access, until the brain can interconnect it with experiences and the iconic memory. This process, mentally, is longer; we only get the singular input, but the brain retains the specifics longer. (Neisser, U. (1967). Cognitive psychology. Englewood Cliffs: Prentice-Hall. ISBN 978-0131396678)

It is precisely because this memory function operates differently, that the auditory and visual learning methods should be both employed to enhance learning overall. Utilizing an auditory learning process can increase knowledge retention, as the information is not only connected to your mental associations through Optic processes, but also through a secondary sensory means. This gives your mind an alternate, and hopefully amplified means to keep and maintain information.

In particular, as the segmentation of what we see is so fast, so broad, and so much more processed than that of what we hear, the benefit of the audiobook over the written book is defined by the amount of retention of the original data. It is true that one can review, and review the material if it is written, but the review functions to reinforce the secondary phase, the iconic memory. When we listen to a source, we capture bigger 'chunks' of the information, and retain it in that original form, without filter or interpretation. Access, therefore is faster, because the data does not have to be re-correlated, reconnected for the thought, the memory to make sense.

Cognition and Memory

In this section, we are going to get down to the brass tacks about memory, and how iconic memory, echo memory, intentional mnemonics and sound theory call play into creating a 'perfect storm' of learning opportunity.

Signal to Noise ratio (what you hear vs. how well you listen)

There is a very well known difference between hearing someone, and listening to them. In like fashion there is a distinct difference between noise (stuff you hear) and signal (stuff you are trying to listen to). When the actual ratio between these gets too off balance, then learning ceases. So it is best to determine what that ratio is for you, and then work to maintain the balance in order to maximize your learning output.

Working Memory vs. Long-term memory vs. Data Retrieval

Functionally, learning consists of three types of memory. Long-term memory is a function of life experiences and situations, wherein we can gather that information when necessary, even if it occurred long ago. Data Retrieval is the memory process we use often, to pull out significant data from our mental records, and fundamentally ensure we can access everything smoothly. Working memory is that part of the brain's function where we discern and make conclusions, and draw up plans and otherwise work with the data from the other two sources. Hearing feeds directly into long-term memory, so it is a matter of training the Data Retrieval processes to rapidly deploy them, and for the Working memory to be able to manipulate them.

Memory and Encryption through sound vs. optics

A recent film had a scene where a data file was encrypted with images and sounds, to make the retrieval a challenge to do without the precise images and specific sound sequence.

In reality, that is how memory works. One finds a piece of information useful, and decides to commit it to memory. We associate that piece of information with the current details, like sounds, sights, feelings, and we log it away with all those details, so we can access it later.

Then, as we require it, we use our minds to determine where it is in memory, connect the dots, as it were, and then pull it back for working memory use.

Chapter 2: Auditory Learning Is A Learnable Skill.

In this chapter, you will learn:

- Key Characteristics of Auditory Learners

- Techniques for using Audio Books to utilize Auditory Learning

What is an Auditory Learner?

In the education processes of today, there are three kinds of learning that have developed, that serve to offer educators more than one way to present information, so that the individual student has more than one way to develop a given skill or knowledge base. Each is incorporating different pathways into the brain of the student, and an awareness of each person's strengths and weaknesses afford a better understanding of how best to reach that particular student with the material they are to learn. Though educators debate, which are more efficient, which work best for a particular subject, most acknowledge it is a combination of all three that has the most effective value.

Visual Learning

As its name implies, the first of these are students who, through visual means such as observation and reading for example retain information easier. They are not great listeners, and may not be able to use touch or hands-on approaches to retain information. For a great majority, this is the strongest method, and that makes sense, as our eyes are our primary sensory input process.

Kinesthetic Learning

Perhaps not as readily identifiable by its long, challenging name, learners who work best by going "hands-on" – using the sensory input of touch – are kinesthetic learners. These are that percentage of the population that likes to take things apart and put them back together, and access learning more easily through practical contact over written or verbal means. They are perhaps the smallest of the functionaries, but there are certain laws and relationships that are best taught through physical contact means.

Auditory Learning

Learning through our ears has long been promoted as important, and the concept of story teller and lecturer as teacher has been a long-time and honored tradition. As we are discussing the merit of audiobooks, here are the major characteristics of an auditory learner. By going over these, we can build an understanding of the value that recorded books can have in improving, and increasing, intellect and learning speed. Thus, here are those characteristics, in no particular order.

Have a knack for ascertaining tone in conversation

An auditory learner is processing the sounds brought in, much as a mining process. The actual information, the vocal inflection of the speaker, the apparent mood and tempo of the message being delivered, all carry information that the hearer then uses to correlate that information into his or her mind.

This can provide insight into the nuances of the words used, and convey considerably deeper meaning than one might be able to get from a written document with the same terms.

Great at writing down what they heard, over what they read

When recalling the lectures and what the speaker had to say specifically, an auditory learner, because he is depending on the longer-band echoic memory, has a greater likelihood of capturing large parts, even verbatim, of information provided in a verbal form. Thus it is that instructors will often write something on the chalkboard (or today, the whiteboard) and recite the same information. Auditory learners have, therefore, a greater likelihood of being a great group-study asset for this reason.

Associate one or more sounds with memory and improves access

Another benefit of the auditory learning process is that external sounds and noises, that in many cases might be a distraction to learning, actually can play into improved memory, by associating the specific interrupting sound with the activity and details of learning. We have all experienced a point in life where a sound triggers a specific memory, or when a song on the radio elicits a bevy of memorable moments in our past. These are the points in time where we have tied that particular sound or song, to a variety of experiences, and when the one memory activates, it triggers recall to the others.

This can be particularly useful when studying for a particular subject, say, and you play a specific song to trigger that memory... so when you are working the project, you can mentally access the music, and with it flood the subject memories as well.

Storytelling, and talking through problems are a result, and why audio books are particularly useful.

Because of the primacy of vision, the echoic memory functions are particularly useful in recalling specific instances that are not directly associated with the details of life. For instance, while you are mowing the lawn, the roar of the mower is acknowledged, but listening to an audiobook while mowing the lawn allows that time, that you certainly could not be reading, to still function in aiding in education.

In fact, the variety of voices that record audiobooks help one get past difficult names, affords stronger context for figuring out words one doesn't know, and provides better cognitive skills by giving the listener a broader range of listening experiences, which in turn makes the hearer a better, more skilled listener and learner.

Techniques to Use with Audiobooks to Amplify Learning

In the search for ways to improve learning skills and velocity, we turn to what we know, before we reach for what we don't know. Lecturing and storytelling are great, but how can Audiobooks

Associate audio passages with visual cues.

Research the history or background of the audiobook to provide visual cues that may be pertinent. Find and familiarize yourself with maps of the locale in which the story of the audiobook takes place, so your mind can connect the places in the recording with special relationships and distances between them. Associate further by, for instance, listening in the out of doors to those parts of the story that occur outside, so the physical condition more clearly matches the story, locking the memories more clearly.

Use Audio books in an environment of meditation

Distractions are always a concern for learning, and choosing to draw information in when in a more placid, meditative state allows brain function to focus in more tightly on the subject matter at hand. Choosing, for instance, an audio book that is slightly above your reading level tends to give you a 'chewier' experience, because the word usage is perhaps a bit more sophisticated, and you have more opportunities to learn to use context to discern meaning, and thereby expand your knowledge of words and phrases. While such doesn't improve spelling skills, it certainly can improve diction and pronunciation.

Listen to Audio books that stimulate emotion

A written book can be emotionally charged, but it is often difficult to discern emotional nuances. An Audiobook, even if the narrator doesn't transmit emotion directly, puts voice to the words, and those words can evince a response nonetheless. By being able to viscerally feel the material, you link in your memory the facts and the words with the emotion, creating a stronger, more lasting, and easier to access thought process.

Associate particular audio books with particular visualizations

If your material has a nautical theme, perhaps viewing paintings of the sea or sailing ships might amplify the experience. If it is set in outer space, perhaps lying under the stars listening will enhance memory and bring the story to remembrance faster and more clearly.

Chapter 3: Audiobooks Have An Education Edge In Options

In this chapter, you will learn:

- Key reasons Audiobooks have an advantage in education

- Key influences that Audio Books can have on alternative education options

Practical Advantages of Audiobooks

As recording processes advanced, so did the options and the access to Audiobooks. Regardless of whether the reading is purely for entertainment or for project research, memorization, or just to pass the time, Audiobooks have certain physical, tangible advantages over traditional paperback or hardbound books. In this chapter, we will look at the benefits that Audiobooks have over their bound brethren.

Mobility

With the positively microscopic audio players of today, there is hardly anywhere that you cannot take an audiobook file with you. Personal sound systems, tablets and smartphone can access mp3 files or other audio formats, and play your audiobook right where you are. Since the feed needs only reach your ears, headphones or the newer, tinier 'earbuds' provide private delivery directly where you want it.

Access to resources to download the files for audiobooks is as simple as accessing the Internet. Publishers create downloadable links on their books' website to make purchase as easy as possible. Public Libraries often have not only the most current, but also classics from every genre and theme. Sites like Amazon and EBay have whole sections devoted to this form of entertainment and education.

Associate intent and context

One of the most challenging aspects of reading any book, is to derive the intent and context of a given passage, in terms of what the author intended for the reader to pick up. Audiobooks provide the reader an opportunity to experience the words aurally, rather than visually, which in most cases provide the listener with insights that make such interpretations easier and reduces the confusion.

Build vocabulary and proper pronunciation

Audiobooks, particularly if they are at or higher reading levels, give you opportunities to discover new words for your vocabulary, and to provide proper pronunciation and inflection, even for words you already know, These benefits will improve your own speaking and listening capabilities, and help develop contextual meanings for words that are unfamiliar to you.

Repeatability due to speed

Another benefit, depending on your own reading speed, can be a quicker process of assimilation of information. Replay and repeat of material from an e-book may be more challenging to attain, but for sheer speed through of content, or the repeat of whole sections of a story, audiobooks make a great choice.

The ability, further, to go over the broader sections does much, too, to avoid the dangers of missing context that can be so problematic when analyzing a written work, word by word.

The Influence Audiobooks have on Alternative Education

Perhaps your learning situation isn't the usual "Monday through Friday" track. Perhaps you are working through a GED or home-school program. Audio books provide some great options for learning, and some great ideas on how to bring those options to practical use. We've gathered a great many of them here.

Offer reading opportunities above the reading level

With as many of the classics of literature, fresh sources and resources are developing every day. Titles from the past, as well as new and entertaining books are being converted for audio listening every day. By being aware of the titles and their relative difficulties, a listener can books that challenge their reading skills and still meet their genre choices and tastes.

Speed of acquisition can be faster

Some people have a gift of learning, and can absorb information at an incredible rate. Others learn speed-reading, and if they are a visual learner, which can mean great advances in their information acquisition. For the rest of us, the ability to take in whole volumes at a conversational pace is a great advantage. Even without speeding up the speed at which the recordings play out, the rate of intake is still several times that of trying to read alone. Further, the effectiveness of transmitting information can be increased as we have talked about earlier, and get significant increases in retention as well.

Develop Critical Listening Skills

It is one thing to hear what someone is saying. It takes a discerning mind to commit information to mind, and by listening and applying the techniques we have talked about, your ability to discern meaning will ramp up considerably.

Increasing Availability

Over time, the number of books you have consumed, and the rate at which you can accumulate that information is also a factor of availability. As time goes by, more and more books become available in audiobook format, and the more likely you can find the ones you are looking for becomes.

Decreasing cost of ownership

Libraries and bookstores that deal in second-hand titles is another resource whose time is at hand. More often, the audiobook titles become available from the publisher, who realizes that it is a great way to introduce the title and bring more fans, further, the very love of reading itself depends on accessibility, so a lower cost audio format will lower the bar to retention even lower.

Audiobooks are a great way to overcome disabilities.

Most of us take for granted the ability to see well, or at least well enough to read. For those that are unable to do so, an Audiobook is a great alternative, where the listener can benefit from the story, but not be punished for not being able to read. Further, and though it is a small percentage, there are still adults in this country and are illiterate. Audiobooks provide both these circumstances with the way to overcome and gain access to the same literature, in a means they can access.

Chapter 4 Audiobooks Allow You To Harness Your Passion

In this chapter, you will learn:

- Why a love for listening equates to a love of learning

- How Audiobooks lead to a love of reading.

Loving to Learn

Advancing skills means you want even more

One thing that seems to be true about learning, is that once you have a little information, you want more. And more. With Audiobooks, the variety and scope of available literature means you can find exactly the right book for any situation.

Doing a little traveling? Pick up a light comedy to keep you engaged and entertained. Got a dark and spooky night ahead? Pick up a thriller, or maybe a gothic romance. In the mood for nostalgia? Find a book that maybe you read long ago, and just want to feel that emotion, that excitement again. Cultivating a romance with reading, with listening, and with learning will enhance your life in ways that go beyond just learning.

Developing the understanding leads to better attention span

The length of time a person can sit and read a book, in the world today, just seems to have dwindled considerably. Being able to pick up a chapter, read it through, and move on to other things helps to insinuate a literary life into your busy schedule. As your interest in the stories increases, so will the time you can afford, since the material becomes a priority. Pretty soon, you will have a greater attention span, and more information will be assimilated faster.

'Burst Listening' is the key to broadening patience and tolerance

With children in particular, the desire to learn can sometimes be hidden or occluded by the desire to play, to interact. So, use a strategy of "burst listening". The learner only takes in a minute or two of information, but then is asked to explain some piece of what they have heard. In essence, reading, listening to an audiobook becomes a game, and the participants learn more, have better access, for the effort.

Vocal Variety leads to listening as entertainment.

As we cover in an upcoming chapter, the narrator makes for the experience of audiobooks. It is totally a personal preference, and to some, the narrator's ability to change his or her voice for a character or situation makes the story more alive. Others are looking for a direct experience so they want a narrator that does NOT make up or change their voice during the recording. The point is, a narrator provides the listener some positive experience from their presentation.

The Benefits of 'Binge Listening'

Just like the recent media attention drawn to people who compulsively watch whole seasons of TV programs end to end, like a food binge, but instead visibly, an audiophile who has a penchant for audiobooks can take on whole series' of works by the same author, or every book that has been narrated by the same person, and so on. The benefits are considerable in comparison to the binge TV watcher; you can listen anywhere, stop and restart without as much confusion, and you can go at your own pace throughout the full series.

Leading to Reading

As an avid reader, audiobooks may seem like an enemy, depleting the ranks of readers across the country. We would disagree, because of the following reasons that we think that a love for knowledge and the tool of audiobooks can and does actually increase the number of people reading overall.

Hearing can lead to reading because of vocabulary.

One way that audiobooks can lead people to read, is by introducing words to their vocabulary. Once it has entered verbally, audibly, there is a desire to know how to spell it and to use it in context. So when the purveyor of the audiobook is finished, even if it isn't initially to read the whole thing, he or she may purchase or borrow the book to discern the vocabulary they have learned.

Accessibility makes for inclusion

Being able to join a community through hearing the books they like in order to fit in is a great way to use Audiobooks. Over time the interest in the topic will inevitably lead to reading the base materials, or expanding the desire to know more about the subject. Ultimately, that will lead to expanded reading as well.

Comprehension is a euphoric

A little knowledge goes a long ways. So it is with the ease of audiobooks. Being able to polish of "Anna Karenina" or "War and Peace" is a literary slog. But the audiobook versions make for lighter fare, and the accomplishment of working through such a beast gives the listener a lift, brings happy and successful feelings and awareness. That, itself, can lead to a desire to read, to catch up, and to fill in the blanks.

Narrators can draw to new genre. Which leads to new learning.

If a narrator appeals to you, you may pick up a very different kind of book if they are voicing it. So goes it for a lot of works. Sam Elliot gets your fancy because of that rugged western voice? Want to hear a book read by its author, and then find others with the same sound? Like a particular accent or dialect? The fascination with the narrators can even lead a person to read, if they then get that 'voice in their head' for their own reading.

Chapter 5: The Power Of Audiobooks Can Be Amplified

We've covered some of the ways that Audiobooks can improve learning by their own merit, but now we are going to demonstrate some means by which Audiobooks can reach even more effectiveness, by merging their effects with the other training methods.

In this chapter, you will learn:

- Associative Power of Auditory Inclusion

- Cognition and Hearing

Memory and Interactive Inclusion

It may seem a simple thing, but being present during the listening process can make huge inroads in learning and increased intelligence. Awareness of the import of memory, and working actively to commit what you hear to the wrinkles in your gray matter can change not only your ability to remember, but your access speed.

Conversational Linking – speak while listening to associate concepts

Some writers place rhetorical questions and the like in their books and even when they don't, consider having a conversation with the writer being quoted. Having a conversation with the story, the characters, or with other imaginary or real hearers can build insight, identify questionable positions within the work, and so much more.

Musical Linking – listen to light music while listening to learn to associate memory

Most people are aware of the link between light classical music with learning and education. The interesting point that makes a real difference is that memorable music tied to data we want to remember is very strong. Consider your own life, and the times you have a particular song cross your mind. Soon, memories of your life will flash by, as the recall processing links all memories associated with that music makes its way past your recall center. In like fashion, play low volume, peppy music to associate with the audiobook passages, to build your access, and your memory speed at the same time.

Subconscious learning – listening while meditating or sleeping.

Our subconscious operates all the time, and is an unfiltered resource. Data goes in all the time and nothing is precluded, so if one achieves a meditative state, or at least is nearing sleep, data is more readily accepted and recorded. For this reason, taking advantage of public transit, long drives, etc. are great places and opportunities to take in an audiobook. Listening as you go to sleep, or listening as you meditate, you give your mind permission to take the information in unimpeded. Staying on track, as you awaken or as you end your meditation, think upon the work you've heard, and commit it to your memory. You will be surprised how much you can recall this way.

Increase Effectiveness by Adding Entertainment

Nobody said learning has to be hard or boring. Adding in levity, fun, excitement only add to the enjoyment of the process, and give the learner a further incentive to participate. Listening to a compelling story is only one way to accomplish this. Below are seven more ideas that will enhance and amplify the experience.

Narration as motivator

Tim Curry. Alex Baldwin. Steve Martin. The list of highly entertaining actors, singers, and entertainers who have become narrators for audiobooks is increasing all the time, and the appeal of the particular reader can even be a further enhancement in terms of incentive to listen. It is more fun to hear a familiar voice reading in many cases than even the author themselves!

Find titles that appeal to your interests

If you like car racing, find books on the subject. Find narrators that are drivers, or stories about drivers. The variety of materials out there mean you can choose what resource you want for information. If in particular you are seeking a certain kind of skill, you can even find 'How To" books in Audiobook form.

Mix it up. Change genres, themes, tempo

Our minds look for variety, for differences, and yet seek patterns. If you are seeking to increase intellect, the easiest way is to take in a wide variety of resources. Maybe this week check into an action adventure story. Next, a dramatic romance. In any case, the skills you learn through audiobooks combine over time, building your acuity and knowledge base.

Listen differently

It may seem odd, but consider HOW you listen, as well as to what. If you generally are following just the facts, consider instead the motive of the author, the inflection of the characters, the purpose of the plot. Shifting your paradigm, your point of view, you get even more information out of the narrative each time you hear it.

Listen for the poetry.

"Words mean things". Sometimes, it isn't so much what we say, but the way we say it. Listen to the words used, and seek out alliteration. Onomatopoeia. Syncopation. There can be artistry designed directly into the emphasis, and you, the reader, can mine the words for so much more than 'just the facts, ma'am'.

Engage the other styles of learning

We have already spoken of it, but it bears repeating. Consider ways to merge the learning processes. Open the book and follow along with the narrator. Notice the pauses, the paragraphs, and the poetry of the terms. Tap out the rhythm of the speaker, the sounds you hear in the terms used. It can be really exciting when you change up what you are looking for when you listen. See what I did there?

Compare and contrast styles within the same work.

Many titles have a variety of narrators. One really fun trick is to listen to one narrator for a chapter, then swap out for another. The perspectives change, the perceptions and inflections on the characters shift. Make the learning process feel alive, and your understanding of it will explode.

Chapter 6: The Narration Can Make A Major Difference

Insert chapter here...

In this chapter, you will learn:

- There is a major difference in Audiobook quality of narration

- Pros and Cons of self-narrated works

Audiobooks and the Narrator

Just as there are thousands upon thousands of titles, there are thousands and thousands of ways an Audiobook can be put together. Below are a few of the more common styles, along with the advantages of each.

Author as Narrator

One of the benefits of being a storyteller, is getting to tell your story. A writer will often gladly read their own work into Audiobook form, initially as an additional revenue stream, and also because it gives a chance for the voices in the story to be heard as the author intended. In such cases, usually the author gives a little 'behind the scenes' kind of additional content, and even may offer suggestions about what was the purpose for this scene or that, or other absolutely invaluable gems of information not found in the written work.

Professional Narration

Some writers don't have a knack for reading or storytelling, and prefer to hire a professional voice to do the reading. These voices are most often quite good, and are rarely improvisational or particularly intuitive. The major advantage is they rarely miss a syllable, and the rendition is most often quite faithful to the written form.

Celebrity Narration

Most often, this form is used when a work has a particular theme or character for whom the Celebrity is a perfect fit. It may also be done when the Celebrity favors the author or perhaps they have a common cause they both believe in. Usually quite good, and the Celebrity adds value simply by lending his or her voice to the project.

Theatric Reading or Audioplay

A great version for gathering perspective on the various characters, the theatric reading or audio play is fundamentally an aural equivalent of a radio play or nonvisual screenplay. Sound effects, vocal variety, and other special sound functions usually accompany the actual text, and even page breaks or chapter breaks have some audial acknowledgement. Some prefer this, particularly for children, but others see the extra sound as 'fluff and nonsense'.

Making your Own Audiobooks

One note that most books on the audiobook process is they usually gloss over one of the most effective learning mechanisms, and that is recording your own audiobooks. You will need some specialized hardware, but going through the process you will further improve your knowledge of the material, simply by being the one to read and record it.

Speaking the words cements them in your mind.

Just as hearing a book read offers additional learning benefit over just reading it, reading a work aloud is another ratchet up the channel of understanding. Learning to vocalize, pronounce, and maintain volume as you record will all reinforce the material as strongly as all the other forms of learning.

Hearing your own voice changes your perspective on the world

Most of us have never had the experience, and once you have, your thoughts about voice, sound and individuality will change. Most of us are displeased with our own voices, and so by hearing yourself, you give yourself the opportunity to begin making changes, to change the way you sound, how clear and understandable you are, and so on.

There is nothing wrong with going through the publication process

Reading someone else's work out loud is a form of performance art, so recording it can actually create a resalable product if you follow all the rule for giving appropriate credits to others. Even if you never get that far, becoming aware of how the system works will give you more education on the entertainment business than you would ever have expected.

Chapter 7: Audiobooks Are Available For So Many Categories

In this chapter, you will learn:

- The wide open field of Audiobooks

- Care and maintenance of your personal audio equipment – your ears.

So, What Kind of Books Can You Listen To?

If I thought it would go without saying, I wouldn't say it. But the truth is, most people don't realize the absolute wealth of reading material that has been converted to Audiobooks, or even that such a marketplace exists. I could literally write for days and days, and not touch the surface of the opportunities.

The categories, however, are a much shorter list, and while I may miss out on a few, the following list should cover most of the bases.

Non-Fiction

Biographies – These include books written about individuals by others or autobiographies, and outline the tales of individuals, struggles, successes and life.

Business - Whether dealing with basics of set-up and finance, or making moves and hostile takeovers, audiobooks give the busy owner or manager a way to keep up his continuing education and situational awareness.

Children's – One of the most entertaining, and largest markets for audiobooks are those that entertain, delight, and educate our little ones. They are a lot of fun, too.

Cooking – Near and dearer still are the diets, recipes, health hints and absolute epicurean delights of this category. Another huge category, with many titles and concepts encompassed by the name.

Essays – From op–ed collections to complete books of perspectives essays tend to include viewpoints, analysis, critiques, and personal & public commentary.

Fantasy – Everyone appreciates escapism, and this broad category covers everything from flights of fancy to eroticism, from twists on reality to full-blown phantasmagoria.

Fiction – Perhaps the biggest category, the works of human imagination from major classics of literature that defy other classification, to simple stories a tall tales, the fiction category generates the lion's share of available audiobooks on the market.

History – Details and descriptions, explanations, and analysis of historic times. From point to point outlines of the course of events, to individual viewpoints at the time, this category has a very broad reach.

Information – Audiobooks from this category include travelogues, guidebooks, educational assistance, and material to support programs in schools and the like.

Inspiration – These are the often dramatic, constantly uplifting and inspiring stories of the people and situations of life who have overcome obstacles to succeed.

Instruction – Step-by-step guides, overviews, how-to books and more fall into this category. If you can build it, make it, or repair it, there's and audiobook for that.

Languages – Learn how to speak in another language by listening and repeating words and phrases.

Law – The information on any aspect, from true crime blotter stories, to legal opinions, to the very words of the laws themselves, and the lawyers that argued for or against them.

Literature – The love of the written word can take a lot of different directions, from scathing critiques to the masterpieces themselves fall in this category.

Performing Arts – Any book on any form of art that you can imagine, from Ballet Dancing to Zumba, and everything in between.

Philosophy and Psychology – What we think, how we think, and why we think it can be tied into these very broad topics.

Politics – Liberal or conservative, Republican or Democrat, there are audiobooks galore that cover this arena.

Religion – Whether living or ancient, dealing with the practices or the pitfalls, all shades of the spiritual side of humanity can be heard in audiobooks.

Science – Every discipline has its writers, and the scientific field has a plethora of professionals that provided hours and hours of audio education and entertainment.

The number of hours of recorded audiobooks already on the market would represent more than a lifetime in hours of recorded knowledge. Take advantage of these learning opportunities as you can. You will be glad you did.

Maintenance and care of your "on-board hardware"

While on the subject of the auditory system and its essential nature, it would be unfortunate if we didn't address some of the tools that can be used to keep our ears working properly.

One such recommended tool for maintaining your hearing would be a properly sized pair of **Earplugs.** Whether you work near loud machinery or do a lot of flying, both consistent droning noise and sharp, staccato burst can do damage to your ears, so by keeping a pair handy, you afford yourself longer use of your ears, and protect them from damage as well.

Believe it or not, whether you have a favorite musical group or sit near a ventilation system that thrums on constantly, it is best if you take periodic **Auditory Breaks** from whatever source. Ears can be damaged by too much of anything, just like any other human activity. Be sure to take ten to fifteen minutes or more, every other hour, to give your listening system a way to self-calibrate from time to time.

Regular checkups, too, are highly recommended. Just like any system, the self-correcting nature causes us to not even be aware when we begin to lose that system's abilities. By keeping up with regular visits, we provide an external check to keep us on the right track.

Common Audiobook Recommendations

There are quite a few things that just about every Audiobook user and publisher would universally agree upon.

Do's

Explore all the possibilities, by searching through the vast catalog of audiobook offerings.

The resources are out there to be perused. The extremely low cost of access means you should be able to get what you are seeking with only a little investigation.

Determine what you are looking for, and don't settle for less.

The vast opportunities means you can find virtually any educational material you want, for a reasonable price. It just takes a little determination.

Use visualization, background music, and meditation techniques to amplify the effect of Audiobook learning.

We all learn differently, and by using more than one set of skills at the same time, you can achieve markedly improved retention and access over any of the processes alone.

Share your experiences with others, and ask a lot of questions along the way.

There is much to learn, and so many people that are on the journey with you are willing and able to lend assistance. In the end, you will find your own success will put you in the role of helping others, and you will then understand the real gift of knowledge is in the expression thereof, not the attainment.

Don'ts

Expect miracles overnight.

Despite the marked improvements most experience in a short while, believing in an instantaneous change in intellect or information retention is unrealistic, and will actually limit your ability to truly commit to the work that developing a learning style takes.

Try to do to too much at once.

Putting the recording at double speed, binge-listening for hours on end, and choosing to attempt too many changes in your learning process at the same time actually can detract from your successes, so instead, work to perfect a few concepts at a time, and record your results. In time, you will see the differences you are trying to achieve come to pass.

Get discouraged.

We all want and desire improvements in our lives. When they don't occur when we would like, we can sometimes reflect that discouragement back as adverse thoughts on our own skills and talents. In the pursuit of improved education, understanding and learning, it is best to measure, not by our own expectations, but by how much more we have in in our minds to access, and how easily that access becomes over time.

Conclusion

Learning and intellect development is a personal process, one that depends on your capability, your willingness, and your objective application of the tools you choose to provide yourself with. Whether you are an auditory learner, a visual learner, or one who learns through touch and interaction, applying even some of the recommendations in this book will automatically improve your learning capability, because to a large extent, any learning process is a combination of the three.

We covered the mechanics of hearing, the barriers to auditory learning, and some skills for improving the outcome of listening to audio books. We went over in some details the benefits of using Audiobooks to not only introduce subjects, but also to review and retain information faster, and more efficiently through them.

The major benefits of an audio book provide the user with easier access, increased availability, and even subconscious application through meditative hearing. Adding to this the associative method, where you put sounds with sights, and the more visceral active listener modes, you take for yourself the means to increase retention, broaden the context, and engage with the speaker to more deeply appreciate the words you are processing.

I trust your journey to discover ways to improve your life through educational advancement and information retrieval through improved memory and increased intellect has been as successful for you as it has been thrilling for me. It is always of great benefit when one can help others achieve their own objectives.

In the end, the love of learning can be increased, enhanced, and amplified through not only listening to learn, but also listening for leisure. With all the benefits you can gather from this book, be sure to share your appreciation with others, and remind them where you found this gem... then go mine for some more!

To hear about Entrepreneur Publishing's new books first (and to be notified when there are free promotions), sign up to their New Release Mailing List.

Finally, if you enjoyed this book, please take the time to share your thoughts and post a review on Amazon. It'd be greatly appreciated!

Thank you and good luck!